OUR CANAL

THE RIDEAU CANAL IN OTTAWA

Peter Conroy

Published by

 GENERAL STORE PUBLISHING HOUSE

Box 28, 1694B Burnstown Road, Burnstown, Ontario, Canada K0J 1G0
Telephone (613) 432-7697 or 1-800-465-6072

ISBN 1-894263-63-4
Printed and bound in Canada

Cover design by Derek McEwen
Colour done by Studio Colour Group
Printing by Custom Printers of Renfrew Ltd.

No part of this book may be reproduced, stored in a retrieval system or transmitted in any form or by any means, without the prior written permission of the publisher or, in case of photocopying or other reprographic copying, a licence from CANCOPY (Canadian Copyright Licensing Agency),6 Adelaide Street East, Suite 900, Toronto, Ontario, M5C 1H6.

Copyright © 2002 by Peter Conroy

National Library of Canada Cataloguing in Publication

Conroy, Peter

    Our canal : the Rideau Canal in Ottawa / Peter Conroy.
ISBN 1-894263-63-4

    1. Rideau Canal (Ont.)--History. 2. Ottawa (Ont.)--Buildings, structures, etc. I. Title.

HE401.R5C66 2002   971.3'84   C2002-902372-6

Sincere thanks go to Nancy, who encouraged this book the longest and loudest. And to Barb Barkley, the Pockets — Anthony, Margaret, Tara and Anne — Heather Lang, Tom Sawyer, Jean Claude Boudreau, Gordon Cullingham, John and Colette Riddle, and Elaine Kenney, all of whom either loaned materials, gave advice, or contributed their time directly towards the production of *Our Canal*.

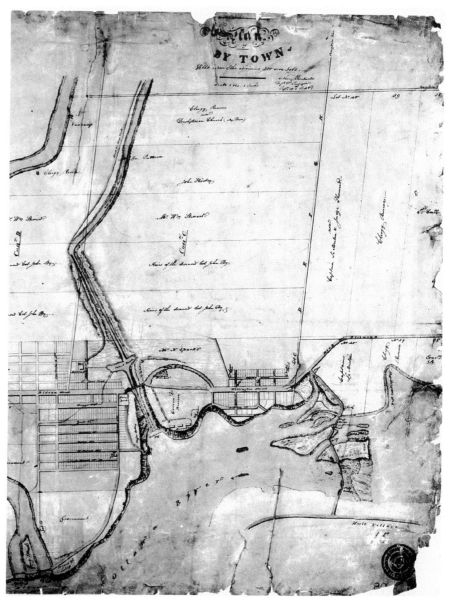

Ottawa in 1840. (National Archives of Canada — detail of NMC21868)

JUST SAY THE WORDS "the canal," and anyone who knows Ottawa knows exactly what you are talking about. It is that seven-kilometre stretch of water (or ice!), bordered by paths and driveways, that snakes through the central part of the city — from the locks at Parliament Hill, to Dow's Lake, and on to Hog's Back Falls. It is an outdoor skating rink, a jogger's dream, a pleasant walk among flowers and trees, a rollerblading track, or a source of boating fun. Visitors to Ottawa may not have time to experience all these pleasures, but even a few days in the capital is enough to make them aware of the beauty of the Rideau Canal, and of what it can offer to enhance their visit.

Of course, the Rideau Canal itself is much more than a piece of Ottawa scenery. It is a 123-mile waterway, stretching south to Kingston on Lake Ontario, an amazing engineering feat in its day, and an important element of Canadian history. So let's take a few minutes to understand why this waterway was built before returning to the subject of this book — the most visible part of the Rideau Canal and the longest man-made stretch of the waterway — the part that runs through the nation's capital.

The Construction of the Rideau Canal

The story of the Rideau Canal begins with the Indian tribes' knowledge of a water route from Lake Ontario up the Cataraqui River, and then down the Rideau to the Ottawa River. It is something of a geological quirk, that a narrow tongue of the Precambrian Shield slips through this area before widening again to form the Adirondack Mountains further south. This narrow stretch of elevated ground allows the Rideau and the Cataraqui Rivers to flow in opposite directions from almost the same source, and thereby to link the Ottawa and the St. Lawrence Rivers.

In 1783, the British army sent a certain Lieutenant French to explore this reported water route, presumably hoping to make use of it. But whatever their hopes and plans, events in Europe and elsewhere seem to have preoccupied the British until after the War of 1812. During that war, when elements in the United States sought to drive the British permanently from "their" continent, it became obvious to the British (though the American enemy took only limited advantage of it) that the St. Lawrence River supply line from Montreal to the vital naval dockyard at Kingston was highly vulnerable. In 1816, the British military command dispatched Lieutenant Jebb to assess the feasibility and the cost of building a canal that would provide an alternate supply route, connecting the Ottawa River and Lake Ontario.

The next move was made by the government of Upper Canada. Alarmed by the economic threat posed by canal building in the United States, especially the Erie Canal, it appointed a commission in 1821 to examine a number of projects aimed at improving transportation within the province. Because the Rideau Canal was considered primarily of military benefit, and perhaps because the Duke of Wellington was known to be personally interested, it was referred to London in 1825 as a project that British authorities might wish to control — and pay for! A special military commission speedily re-examined the defence rationale, the cost estimates, and also the possibility of a co-operative funding arrangement with the government of Upper Canada. Perhaps not surprisingly, it recommended that the British government proceed on its own. Fear of a military attack from the south still running high, and the Duke of Wellington prominent in the British government (prime minister from 1828 to 1830), authority was immediately granted and Lieutenant Colonel John By of the Royal Engineers was selected for the task. The Rideau Canal was built between 1826 and 1832.

Although designated a military waterway until 1853, the Rideau Canal never served the military purpose for which it was so urgently commissioned. However, it did serve very well in a commercial capacity until rail transportation gained the advantage. Since 1920, the Rideau Canal has been almost entirely a recreational waterway.

Ottawa's Origins

Colonel By sailed up the Ottawa River to Wright's Town, a thriving community founded in 1800 by Philemon Wright (renamed Hull in 1875, but now known as Gatineau). By 1820, Wright's Town, or Wrightville as it was often informally called, had a population of 703 and had based its prosperity on the squared timber needed by the British Navy. In stark contrast, there were only six houses on the Ottawa side of the river in September 1826, and almost impenetrable bush. So impenetrable was it, that the first survey for the beginning of the canal had to be postponed until the ground had frozen, so as to permit the surveyors to move more easily. Even then it took them five days to get from the Ottawa River to Dow's Great Swamp — not even the length of what is now the Ottawa portion of the canal.

The building of the canal created a community called Bytown on the south bank of the Ottawa, and positioned the hub of that community around the flight of eight locks needed to lift vessels up almost 80 feet. Colonel By built his residence on the east side of these locks, in what is now Major's Hill Park. Two companies of Royal Sappers and Miners were stationed on the west side, on what is now Parliament Hill, but was then, inevitably, known as Barracks Hill. Today, this area at the head of the locks remains the focal point of the city of Ottawa.

From its early beginnings as a community of canal builders and those who could profit from the canal, Bytown grew and consolidated its position as a recreational and commercial hub, particularly for the raftsmen of the Ottawa Valley lumber industry. An essential part of this commercial activity was a turning basin for ships at the spot where the Mackenzie King Bridge now stands. The needs of the new railway station eliminated the east side of this basin in 1908, but the western portion remained in place until the late 1920s. It is rather interesting to note how the canal's status as a commercial route was reflected in the building of a railway station at the head of the locks, with the tracks themselves paralleling the canal. When you think about it, even the new railway yard on the east side was just a more modern form of turning basin.

Selecting a Capital

On January 1, 1855, Bytown gained city status and changed its name to Ottawa. At this time, the long-standing dispute between Upper and Lower Canada over the location of the capital of the new united provinces of Canada was reaching its climax. Kingston had initially been nominated, but after three years of vigorous debate the legislature chose Montreal. Then Montreal blotted its copybook when riots in 1849, generated by the Rebellion Losses Bill, led to the burning down of the houses of parliament. At that point, parliament decided to alternate sessions between Quebec City and Toronto. In 1856, tiring of this impractical compromise, all four of the above cities were invited to submit proposals to become the capital; Hamilton and Ottawa decided to enter unsolicited bids. In 1857, Queen Victoria was asked to make the difficult and highly controversial final choice. Ottawa won primarily because it was not one of the original four competitors, and because it lay right on the border between Canada East and West. It also offered a beautiful site for the new parliament buildings. But among the factors that favoured Ottawa, a military argument was again made that Ottawa was more easily protected from the Americans than any site on the St Lawrence River, such as Quebec City, Montreal or Kingston, and that it could be provisioned more securely, from either Quebec, or Kingston by means of the canal.

The Route through Ottawa

Familiar as we are with the focal point of Ottawa provided by the set of eight locks, the Château Laurier Hotel, and Confederation Square with its War Memorial, it might have been very different! All the early travellers who landed on the south bank of the Ottawa River did so at what was known as Richmond Landing, a spot just below the Chaudière Falls. This would have been the logical place to begin the canal. Not only would it have been much closer to Dow's Great Swamp, but it would have allowed a useful spur to enter the Ottawa River above the falls. The story goes that Captain John LeBreton had purchased this land speculatively several years earlier, on the basis of confidential information about the government's plans for a warehouse, and had so outraged the governor, Lord Dalhousie, with the exorbitant price he demanded, that this site was automatically disqualified when planning began for the canal. Imagine what the centre of Ottawa would have looked like had the canal begun at Richmond Landing, crossed LeBreton flats, and run to Dow's Lake somewhere in the vicinity of Preston Street.

Instead, Lord Dalhousie quietly purchased the land above Sleigh Bay, correctly anticipating that it would prove to be the next most logical place to begin the canal. The name Sleigh Bay name came about when one of Philemon Wright's sons used the spot for his wedding ceremony — the guests observing the marriage from a semicircle formed by their sleighs. The reason for performing the marriage in this wilderness was less romantic than one might expect: The justice of the peace selected came from Perth, and was therefore legally obliged to perform the ceremony somewhere on the Upper Canada side of the river. Once the site was approved by Colonel By for the canal it became known as Entrance Bay.

A flight of eight locks was built here, no easy task through unstable clay, permitting a lift of about 80 feet to the top of the cliff above the river. Not far from the cliff, and perhaps a reason for choosing Sleigh Bay, Colonel By used a 12-acre beaver meadow to allow the relatively easy excavation of the large turning basin for the ships that would use the canal.

Beyond the beaver meadow the work again became more difficult. Colonel By and his men were obliged to cut a deep trench, about three-quarters of a mile directly south through the same clay, in order to reach a natural depression that contained a stream. The construction records detail many slides, and much danger and death encountered during this work. For many years after, this first straight section of the canal beyond the locks was known simply as the Deep Cut.

The next part of the canal, starting at the corner after the Deep Cut, used a natural depression and marshy area that drained off towards the Rideau River. Several of these streams to the southeast had to be dammed, so as to fill the depression with sufficient water for navigation. These dams backed up what became known as Patterson's Creek and Brown's Inlet, and up a third inlet into the Exhibition Grounds which has since disappeared.

Beyond Brown's Inlet, Colonel By encountered a piece of high ground, known in those days as the Mountains of Nepean. This term sounds pretentious to most Ottawa residents today. But when you consider the height above the canal of the last portion of Echo Drive near Bank Street, and how that ridge parallels the canal before swinging north across it to become the high ground on the east of Dow's Lake, a better picture can be formed in the mind's eye.

A second deep cut was made, running west through a notch in these mountains towards Dow's Great Swamp, and this very narrow part of the canal came to be called the Notch of the Mountain.

To permit navigation, Dow's Great Swamp was turned into Dow's Lake by means of two dams. The main one was on the south side, now the roadbed for Colonel By Drive as it runs past Carleton University up to Hog's Back Falls. The second, known as the St. Louis Dam, lay half a mile north and is no longer evident.

The last section of the artificial waterway made use of a stream bed that ran down from the Rideau River at Hog's Back to Dow's Great Swamp. At Hog's Back, an impressive 45-foot-high dam was built (17 feet higher than any previous dam in North America, though immediately surpassed by the Jones Falls Dam once the Rideau Waterway was completed). The Hog's Back Dam was needed to flood a difficult portion of the Rideau River, and to create a reservoir to feed the entire artificial waterway down to Entrance Bay. Halfway along this last section, the two locks at Hartwell's were built, raising the water level 21.5 feet above Dow's Lake. At Hog's Back itself, two more locks climbed the last 13.5 feet, to the level of the Rideau River.

The Improvement of Ottawa

Sir Wilfrid Laurier, prime minister of Canada from 1896 to 1911, deplored the ugliness of Ottawa and resolved to do something about it. In 1899, he convinced parliament to create the Ottawa Improvement Commission (later to become the Federal District Commission, and today known as the National Capital Commission). This voluntary body was charged with improving and beautifying Ottawa, in co-operation with the city itself.

The Ottawa Improvement Commission gave early priority to cleaning up the commercial aspects of the Rideau Canal, and acquiring land for a scenic driveway along its western bank, initially only as far as Patterson's Creek. Noticing that the flooded low-lying area at Patterson's Creek offered tremendous potential for beautification, the commission gave great prominence at the time to Central Park, stretching west from the canal at Patterson's Creek all the way past Bank Street. Here, a multitude of paths and flower beds were laid out, and a heavily decorated, pseudo-oriental summerhouse was built on an island. Within two years the Rideau Canal Driveway extended along the western bank of the canal all the way to Dow's Lake, bordered by flower beds, trees, shrubs, footpaths, and a variety of rustic wooden bridges, summer houses, and benches. These "rustic" improvements attracted some public scorn, which may in the end have contributed to their eventual disappearance.

Because the stables at the Lansdowne Park Exhibition Grounds sat right at the edge of the canal, this initial driveway veered west at Fifth Avenue, running through an elaborate entrance archway into Lansdowne Park, directly across the park grounds, and then across Bank Street and Brown's Inlet before rejoining the canal. Only in 1926 were the stables removed, and the Central Canada Exhibition permitted the commission to put the portion of the Driveway that ran through the Exhibition Grounds right at the edge of the canal.

For similar reasons a causeway was built across Dow's Lake in 1903, taking the Driveway to the Arboretum at the Central Experimental Farm as planned, but avoiding the J.R. Booth "Fraserfield" lumberyard and railway tracks situated at the marshy northeastern corner of the lake. This causeway began at Lakeside Avenue and ran diagonally to the corner of the lake nearest Preston Street. In 1927, following Booth's death, the newly empowered Federal District Commission was finally able to construct a driveway entirely along the edge of the lake, and to remove the causeway. The retaining walls around Dow's Lake date from that time, although those along the canal itself were built in stages, beginning in 1911.

The western portion of the turning basin was also removed in 1927, making more room for what is now Confederation Park. The rustic structures, including the entrance to Lansdowne Park and the summer house at Patterson's Creek, disappeared without a trace during the Depression of the 1930s and World War II. Funding levels must have been constrained during this period, but just as importantly, the attention of the commission had moved on to other priorities, such as the building of Echo and Island Park Drives.

The railway lines in Ottawa provide an interesting footnote to the canal story — a problem that eventually became an opportunity. Welcomed initially as a modern and essential form of transportation, 11 railway lines, requiring 150 level crossings, were allowed to crisscross the heart of the city. From 1915 on, a consistent theme among recommendations to the Ottawa Improvement Commission and its successor became the need to solve the traffic flow and aesthetic problems generated by the railway lines. It was 1927 before any progress occurred. The removal of the J.R. Booth railway yard at Dow's Lake permitted changes to the shoreline of the lake and the creation of a park, which today is the focal point of the annual Tulip Festival. After that, matters were stalled again until 1950. But in the end, it was railway rights of way that provided for Colonel By Drive on the east bank of the canal, the Queensway through the centre of Ottawa, and for a number of other parks, parkways and transitways throughout the National Capital Region.

Recreation

Even during its military and commercial periods, the citizens of Bytown and Ottawa always recognized the recreational possibilities of the canal. Winter fairs in Ottawa date back to the early days of the city, and both the frozen Ottawa River and the Rideau Canal provided wide open spaces for races of one kind or another, and for skating. The turning basin was being used as a public skating rink in 1876.

Swimming in the canal has been prohibited since at least the 1950s, but before that it was common. There is even record of an 1893 complaint about nude bathing.

Boating on the canal was a natural leisure attraction, and boathouses were common along its banks, until the Ottawa Improvement Commission began to prohibit them for aesthetic reasons. For many years, the Rideau Aquatic Club (renamed the Rideau Canoe Club in 1902) had its headquarters on the canal, near Fifth Avenue and the ornate entrance to Lansdowne Park. Dow's Lake has always provided a handy — and in time also an attractive — expanse of water for boating and sailing. From 1947 until 1988, the Civil Service Recreational Association operated a sailing club on the lake.

Today, through the efforts of the National Capital Commission and Parks Canada, the Rideau Canal is firmly established in our minds as a place to play and relax. Paved paths run the length of both sides of the waterway, shared by walkers, joggers, and in-line skaters. A beautiful arboretum, created in 1888 as part of the Canadian government's Central Experimental Farm, runs down from the high ground on the west side of Dow's Lake to the water's edge. The large pavilion at Dow's Lake includes a restaurant and rental outlets for in-line skates, pedal boats, rowboats, and the canoes that often make their way a considerable distance up the canal towards the entrance locks. Sometimes racing shells are seen, as rowers use Dow's Lake or the calm, straight stretches of the canal to practise their technique. It is not unusual to find the odd fisherman leaning over the canal railing, enjoying a bit of catch and release in the warm sun.

For each of the past 30 years, the full length of the canal has been cleared of snow and maintained as a public skating rink for as long as the winter weather will permit. In February, the National Capital Commission organizes a popular two-week winter fair known as Winterlude, with the canal as the primary ice-covered location for many of the events.

Since 1951, the Tulip Festival has been held during the last two weeks of May, attracting visitors from across Canada and around the world. Again the canal provides a venue for key events. Dow's Lake Park is in its splendour, filled with thousands of tulips sent to Ottawa each year by the people of Holland in gratitude for having welcomed and cared for Princess Juliana, their future queen, and her family during the Second World War.

Our Canal

The Rideau Canal fathered Bytown and helped it to survive and prosper. Without the Rideau Canal there might have been no Ottawa, and the capital of Canada would probably have been placed elsewhere. Today the canal in Ottawa is appreciated as a focal point in the aesthetic and recreational life of the city, and as a feature of the national heritage that all Canadians are proud of. These pictures of the canal are intended as a celebration of what the canal was, and what it has become, for all Canadians.

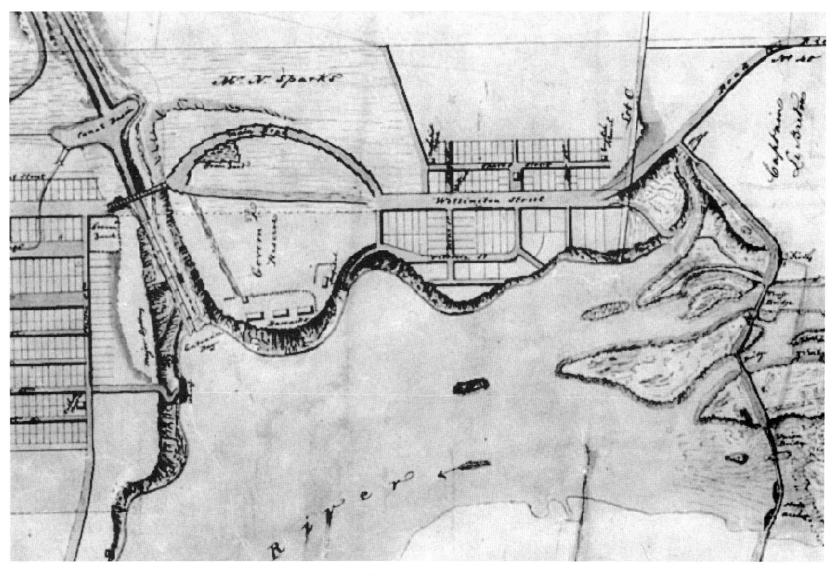

This 1840 map shows the canal, the turning basin, Barracks Hill, the separate settlements known as Lowertown and Uppertown and, on the far right, part of the controversial LeBreton property.

(National Archives of Canada — detail of NMC21868)

1908 — half a turning basin and a bridge at Cartier Street. The Rideau Canal Driveway goes through the Exhibition Grounds, and over the Dow's Lake causeway. J.R. Booth's rail lines sit east of Dow's Lake. (National Archives of Canada — detail of NMC22622)

The locks and Major's Hill in 1861. The lockmaster's house disappeared when Major's Hill became a park in 1876. Further down the locks stands the Royal Engineer's Office, eventually lost to the railway tracks.

(E. Spencer/National Archives of Canada/C34001)

From Parliament Hill — the locks circa 1860. The lumber in the foreground was being used to construct the new parliament buildings. The eastern side of the turning basin can also be seen.

(S. McLaughlin/National Archives of Canada/C610 cropped)

1922 views from the top and bottom of the toboggan chute at the Château Laurier Hotel. For 10 cents, you could slide all the way out onto the Ottawa River.

(W.J. Topley/National Archives of Canada/PA12648 & PA12630)

1900 — a group of snowshoers go down the main locks to the river — the hard way. (Samuel J. Jarvis/National Archives of Canada/PA25007)

The upper locks and the Royal Engineer's Office are visible beyond Sappers Bridge in this 1873 picture, taken during the construction of the Dufferin Bridge.

(W.J. Topley/National Archives of Canada/PA9305)

Dufferin Bridge being built in 1874, forming a triangle with the existing Sappers Bridge. In 1912 both bridges were demolished in favour of Connaught Place, today part of Confederation Square.

(National Archives of Canada/C493)

The Sappers and Dufferin Bridges, as they appeared in 1898. The Ottawa Post Office, on the left, was demolished in 1937 to make way for the National War Memorial. The lockmaster's office still stands at the top of the locks. (National Archives of Canada/C 3778)

1907 — a cyclist watches the trains. In the background stands the Alexandra Bridge, or Interprovincial Bridge, originally built for trains in 1901.
(Woodruff/National Archives of Canada/C8500)

In the early 1920s, the commercial use of the canal was still much in evidence. (National Archives of Canada/PA25000)

A dory sits at the edge of the canal in 1907, next to what is now the outdoor patio of the National Arts Centre restaurant, Le Café. Central Station is on the right and Sappers Bridge is in the background.

(National Archives of Canada/C5276)

A 1909 view of the canal turning basin and train tracks. This picture shows the western wing of the basin, the eastern part having been very recently filled in to accommodate the tracks.

(National Archives of Canada/C5277)

Late 1920s — before Confederation Square and the National Arts Centre.
(National Archives of Canada/PA126433)

1907 — reflections on the Laurier Avenue Bridge.
(National Archives of Canada/C8488)

North of the Laurier Bridge in 1929. The canal turning basin had just been removed and the cleanup is evident.

(National Archives of Canada/PA48161)

Dog Derby Day on the canal — 1931. This event was still being held in the 1950s.

(R.J.C. Fabry/National Archives of Canada/PA148966)

Crowds line the Deep Cut on the first day of the Dog Derby — February 9, 1955. (Andrews-Newton/City of Ottawa Archives/CA7352)

Looking north along the new Government Driveway in 1903. Two years earlier when the Maria Street Bridge (now the Laurier Bridge) was built, the *Ottawa Citizen* deemed it strong enough to support an elephant.

(City of Ottawa Archives/CA1352)

In 1903, this pagoda stood at the intersection of Somerset Street and the Government Driveway. (City of Ottawa Archives/CA18932)

Boat traffic in the vicinity of Delaware Avenue, east of what is now the Pretoria Bridge. By 1911, the new Rideau Canal Driveway, on the west side of the canal, was attracting development.

(W.J. Topley/National Archives of Canada/PA9432)

The Driveway near Delaware Avenue in 1911. In sharp contrast with today's philosophy, the small sign says, "No bicycles allowed on walks."
(W.J. Topley/National Archives of Canada/PA9916)

1891 — salvaging a small freight locomotive that was driven off the swing bridge at the point where the Queensway now crosses the canal. The locomotive's bell sat on a nearby church until the late 1970s.

(James Ballantyne/National Archives of Canada/PA127267)

The Queensway now crosses the canal at this spot where the railway tracks did in 1910. Behind the train is the Victoria Memorial Museum, now the Canadian Museum of Nature.

(W.J. Topley/National Archives of Canada/PA9936)

The Pretoria Avenue Bridge as it looked in the 1920s from The Driveway. A railway car and level crossing can be seen in the centre background where the Queensway embankment now stands.

(National Archives of Canada/PA34317)

In the late 1950's, a railway yard still sat at the foot of Elgin Street where the Queensway would soon run.

(Alex Onoszko/City of Ottawa Archives/CA8167)

The heavily decorated, pseudo-oriental summer house was the crown jewel of the Ottawa Improvement Commission's early work. It stood in Central Park on the island in Patterson's Creek.

(W.J. Topley/National Archives of Canada/PA9947)

Cedar Lodge — the summer house in Central Park. This rustic style, using the natural shapes of tree branches for railings and furniture, was popular for a time.

(Special Report of the Ottawa Improvement Commission — 1913/National Archives of Canada/C10965)

Central Park in 1911 — looking east from Bank Street. Clemow Avenue, originally intended to be a second parkway, was banked up, damming off the creek from the western end of the park.

(W.J. Topley/National Archives of Canada/PA10130)

The bridge at Patterson's Creek, circa 1910, looking towards Lansdowne Park. Boathouses still line both sides of the canal.

(W.J. Topley/National Archives of Canada/PA9931)

Looking south from Third Avenue. The barely seen structure at right is the ornate entrance to Lansdowne Park. (City of Ottawa Archives/CA0257)

This lily pond, near Fifth Avenue, was created as part of the original Ottawa Improvement Commission landscaping. The pond is still there, although the wooden bridge is long gone, as is the Rideau Canoe Club. (National Archives of Canada/PA34085)

The Rideau Canoe Club stood at Fifth Avenue near the rustic wooden entrance to Lansdowne Park until the 1930s.

(W.J. Topley/National Archives of Canada/PA9041)

The decorative entrance to Lansdowne Park on the Rideau Canal Driveway is reputed to have contained 3,000 varieties of Canadian wood. A corner of the Rideau Canoe Club can be seen in the background.

(W.J. Topley/National Archives of Canada/PA9942)

The five-year old Aberdeen Pavilion in 1903, looking almost as sharp as it does today. This view is from the Bank Street side of Lansdowne Park, looking east towards the canal.

(W.J. Topley/National Archives of Canada/detail of PA9125)

Top: Henry's Boathouse on the south shore opposite Lansdowne Park — 1920s.
Bottom: Pagoda, west of Bank Street — 1903.

(City of Ottawa Archives/CA15373)
(City of Ottawa Archives/CA18713)

The new Bank Street Bridge, circa 1913. Until 1927, a fence around the Exhibition Grounds extended to the water's edge, forcing The Driveway through Lansdowne Park and across Bank Street and Brown's Inlet.
(H.J. Woodside/National Archives of Canada/PA16798)

Boathouses along the canal, like these seen from the old Bank Street bridge, were prohibited by the Ottawa Improvement Commission after 1912.

(W.J. Topley/National Archives of Canada/PA9197)

The Brown's Inlet area from the Bank Street Bridge in the 1920s. The south bank still remains undeveloped, although there are rail tracks there somewhere.

(National Archives of Canada/PA34383)

Brown's Inlet in the 1920s, from Wilton Crescent. (National Archives of Canada/PA34387)

1903 — just west of Brown's Inlet on the Government Driveway — the knolls on which the arbour and other rustic pieces were placed.
(City of Ottawa Archives/CA1354)

The rustic arbour near Bank Street, circa 1911. In the background, Brown's Inlet curves towards where the new Bank Street Bridge was built in 1912.

(W.J. Topley/National Archives of Canada/detail of PA9952)

The Bank Street Bridge in 1948, from a leafier Driveway.
(Greber Report/City of Ottawa Archives/CA4055)

The rustic arbour, from across Brown's Inlet. One of these elms still stands.
(W.J.Topley/National Archives of Canada/PA9697)

The rustic arbour as seen from the edge of Brown's Inlet. A horse-drawn buggy can be seen passing between the knolls on the Government Driveway itself. The boathouses on the left are on the south bank of the canal.

(City of Ottawa Archives/CA0255)

Rustic benches and stairs, typical of the original landscaping, on the knolls just west of Brown's Inlet, heading into "the notch in the mountain."

(W.J. Topley/National Archives of Canada/PA9953)

Boaters in The Notch in 1898, before the Ottawa Improvement Commission began its work. (City of Ottawa Archives/CA1602)

The Bronson Avenue swing bridge in 1911. Bronson only became an important street and acquired its current bridge when government buildings were placed at Confederation Heights in the late 1950s.

(W.J. Topley/National Archives of Canada/PA9969)

The Driveway at the southeast corner of Dow's Lake in the 1920s.

(National Archives of Canada/PA34392)

From the Arboretum circa 1905, the new causeway across Dow's Lake. The causeway remained until 1927, when J. R. Booth's lumberyard was removed from the northeast corner of the lake.
(City of Ottawa Archives/CA2526)

Two gentlemen survey the Arboretum in 1898, from the vicinity of the current lookout. The railway bridge at the exit from Dow's Lake was replaced by a tunnel in 1968.

(City of Ottawa Archives/CA1486)

Hog's Back Falls in 1892.

(W.J. Topley/National Archives of Canada/PA33932)

Entrance Bay — the main locks sit below the Château Laurier Hotel, built in 1912.

(Peter Conroy)

To the amusement of some tourists, the original mechanisms, known as crabs, are still used to open and close the lock gates.   (Peter Conroy)

The Historical Society of Ottawa runs the Bytown Museum out of Colonel By's original commissariat building. (Peter Conroy)

Skaters, young and old, throng the Deep Cut on a lovely winter afternoon. (Francine Héroux)

Where once a dory sat, the National Arts Centre restaurant offers very fine food, all of distinctly Canadian origin, right beside the canal.
(Peter Conroy)

By dusk, on Canada Day, every mooring spot just above the locks is certain to be occupied, as crowds gather for the fireworks extravaganza over Parliament Hill.
(Photo:NCC/CCN)

For recreation, exercise, or just plain getting to work, the canal and its pathways are a pleasant experience, in any season. (Peter Conroy)

The Château Laurier Hotel dominates the evening skyline along Colonel By Drive. No wonder Ottawa natives sometimes refer to the city as "Disneyland on the Rideau."
(Peter Conroy)

The lights near the southern end of the Deep Cut reflect off the surface of the canal. (Peter Conroy)

A pair of cyclists pause to enjoy the early morning sun and serenity.

(Photo:NCC/CCN)

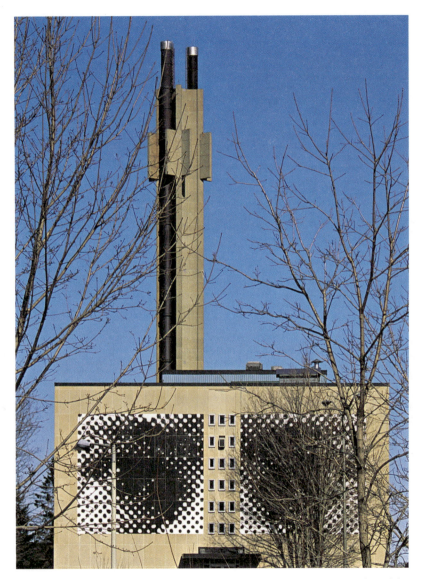

Not everyone notices the "eyes" of Ottawa University's Macdonald Hall, peering down over Colonel By Drive. (Peter Conroy)

A well co-ordinated jogger approaches the end of the Deep Cut. (Peter Conroy)

Now the Portuguese Community Centre, this former church was the recipient of the 1891 locomotive's bell. (Peter Conroy)

Paths along the canal don't sport signs prohibiting cycling these days. Now they are called bike paths. (Photo:NCC/CCN)

Paul's Boat Lines operates popular, guided tours on the Rideau Canal and the Ottawa River. (Peter Conroy)

A solitary kayak makes its way up towards the Pretoria Street Bridge on a calm fall morning.

(Peter Conroy)

This house near Central Park cuts a dashing figure among the numerous heritage buildings beside the canal. (Peter Conroy)

An old willow next to the Patterson Creek bridge gleams in the sun after an ice storm. (Peter Conroy)

An ornate wooden summer house known as Cedar Lodge once sat on this island in Patterson Creek. Unfortunately, it wasn't maintained during the Depression and the Second World War, and it has disappeared without a trace.

(Peter Conroy)

Once the pride of the Ottawa Improvement Commission's early efforts, the park around Patterson's Creek is a tranquil spot. Its original name, Central Park, is no longer familiar to most Ottawa residents.
(Peter Conroy)

The Aberdeen Pavilion, now preserved and restored, has sat near the edge of the canal since 1898. In 1902 and 1904, Stanley Cup hockey games were played here.
(Peter Conroy)

The Aberdeen Pavilion, known sometimes as the Cattle Castle, silhouetted by a glorious sunset. (Peter Conroy)

A different mood — a foggy Saturday morning along Colonel By Drive. (Peter Conroy)

The Central Canada Exhibition has been held in Lansdowne Park since 1886. The ferris wheel has been a familiar annual sight for almost as long.
(Peter Conroy)

Colonel By Drive in a snowstorm. (Peter Conroy)

A young green heron learns to fly near Patterson Creek. A cleaner canal has attracted more wildlife. (Peter Conroy)

The Bank Street Bridge at dusk, before its most recent facelift. (Peter Conroy)

Summer and winter scenes of the magnificent elm tree that shaded the rustic arbour. It still stands next to Brown's Inlet. (Peter Conroy)

Some Ottawa residents are fortunate to be able to enjoy Brown's Inlet as an extension of their garden. (Peter Conroy)

Old steps that lead nowhere are all that remain of the original Ottawa Improvement Commission landscaping. (Peter Conroy)

A car climbs the Mountains of Nepean. (Peter Conroy)

The pleasures of solitude and catch and release, where the canal meets Dow's Lake. (Peter Conroy)

Two runners skirt Dow's Lake in the fog. (Peter Conroy)

The willow trees along Dow's Lake are particularly eye-catching in the early spring.

(Peter Conroy)

Remnants of the old causeway that crossed Dow's Lake can still be seen when the water is low. A causeway was needed until 1927 because the driveway could not run through J.R. Booth's lumberyard at the northeast corner of Dow's Lake. (Peter Conroy)

A music school provides a concert in Dow's Lake Park during the annual Tulip Festival. (Peter Conroy)

The Senator on Dow's Lake. Several boats are available for trips on the Ottawa River or the Rideau Canal. (Peter Conroy)

This pavilion at Dow's Lake replaced the old boathouse in 1985. If you want to rent a boat or in-line skates, or if you want a meal or a drink, this is the place to go.

(Peter Conroy)

Looking north across Dow's Lake from the original dam wall, now Colonel By Drive. (Peter Conroy)

The Winterlude snow sculpture contest on the the frozen surface of Dow's Lake. (Photo:NCC/CCN)

The Arboretum on the west side of Dow's Lake — a popular spot for a walk on a summer evening. (Peter Conroy)

The large field next to Carleton University often serves as a launching pad for sending colourful hot air balloons up over the canal and the Arboretum.
(Peter Conroy)

Carleton University's Dunton Tower, from the Arboretum.

(Peter Conroy)

Looking north back towards the city from the lock at Hog's Back, the last section of the Rideau Canal in Ottawa. (Photo:NCC/CCN)

About the Author

Peter Conroy was born in Montreal and now lives in Ottawa with his wife Nancy. He has a BComm from the University of British Columbia and an MA (Public Administration) from Carleton University. Peter is a member of the Club de photographie polarisé de l'Outaouais. He has lived in Ottawa for 34 years and photographed the Rideau Canal for 25 of them. His love for the beauty of the canal blossomed during a period when he lived just off Echo Drive, and cycled, skated, walked his dog, and photographed to his heart's content. After retiring from an executive career in the Canadian public service, he started looking at old photos of the canal in the National Archives of Canada and saw things he had never heard of. The photos raised a host of questions, the answers to which form the narrative of this book.